ATTEMPTS AT A NATURAL ARRANGEMENT OF BIRDS

BY

ALFRED RUSSEL WALLACE

British Library Cataloguing-in-Publication Data
A catalogue record for this book is available from the
British Library

Alfred Russel Wallace

Alfred Russel Wallace was born on 8th January 1823 in the village of Llanbadoc, in Monmouthshire, Wales.

At the age of five, Wallace's family moved to Hertford where he later enrolled at Hertford Grammar School. He was educated there until financial difficulties forced his family to withdraw him in 1836. He then boarded with his older brother John before becoming an apprentice to his eldest brother, William, a surveyor. He worked for William for six years until the business declined due to difficult economic conditions.

After a brief period of unemployment, he was hired as a master at the Collegiate School in Leicester to teach drawing, map-making, and surveying. During this time he met the entomologist Henry Bates who inspired Wallace to begin collecting insects. He and bates continued exchanging letters after Wallace left teaching to pursue his surveying career. They corresponded on prominent works of the time such as Charles Darwin's *The Voyage of the Beagle* (1839) and Robert Chamber's *Vestiges of the Natural History of Creation* (1844).

Wallace was inspired by the travelling naturalists of the day and decided to begin his exploration career collecting specimens in the Amazon rainforest. He explored the Rio Negra for four years, making notes on the peoples and languages he encountered as well as the geography, flora, and fauna. On his return voyage his ship, Helen, caught fire and he and the crew were stranded for ten days before being picked up by the Jordeson, a brig travelling from Cuba to London. All of his specimens aboard Helen had been lost.

After a brief stay in England he embarked on a journey to the Malay Archipelago (now Singapore, Malaysia, and Indonesia). During this eight year period he collected more than 126,000 specimens, several thousand of which represented new species to science. While travelling, Wallace refined his thoughts about evolution and in 1858 he outlined his theory of natural selection in an article he sent to Charles Darwin. This was published in the same year along with Darwin's own theory. Wallace eventually published an account of his travels *The Malay Archipelago* in 1869, and it became one of the most popular books of scientific exploration in the 19th century.

Upon his return to England, in 1862, Wallace became a staunch defender of Darwin's landmark work *On the Origin of Species* (1859). He wrote responses to those critical of the theory of natural selection, including 'Remarks on the Rev. S. Haughton's Paper on the Bee's Cell, And on the Origin of Species' (1863) and 'Creation by Law' (1867). The former of these was particularly pleasing to Darwin. Wallace also published important papers such as 'The Origin of Human Races and the Antiquity of Man Deduced from the Theory of 'Natural Selection" (1864) and books, including the much cited *Darwinism* (1889).

Wallace made a huge contribution to the natural sciences and he will continue to be remembered as one of the key figures in the development of evolutionary theory.

Wallace died on 7th November 1913 at the age of 90. He is buried in a small cemetery at Broadstone, Dorset, England.

ATTEMPTS AT A NATURAL ARRANGEMENT OF BIRDS (1856)

If we examine the varied form and structure of Birds with a view to their natural arrangement, we see as it were intuitively, that certain well-marked groups exist, which can be distinctly separated from the class, can be easily defined, and will contain species which are more nearly related to each other than to any other birds. Such are the Swimmers and Waders, which together may be called Water Birds, and of the propriety of the separation of which from the whole of the Land birds there has never been a difference of opinion among naturalists. Again, among land birds the Accipitres or Raptores of naturalists, containing the Hawks, Vultures, and Owls, form a well-marked group, all the members of which are undeniably related among themselves, but are separated as it were by a chasm from all other birds; for we consider the supposed affinity of the Owls with the Goatsuckers to be quite incorrect; those birds resembling each other only in a few unimportant particulars, while in all essential points of structure they widely differ. A third group which can also be readily distinguished and separated from the rest, is that of the Rasores or Gallinaceæ. The Pigeons are generally included in this order; but in that case a definition of Gallinaceæ becomes impossible, as so many of their most marked peculiarities do not exist in the Columbæ. It is however extraordinary, that though the Pigeons possess more characters which connect them with Perching birds than with Rasores, yet it is more easy to conceive their connexion by intermediate links with the latter than with the former; for it has never yet been pointed out what particular family or genus of Perching birds makes the least approach to a Pigeon. We therefore conceive that the Columbæ should form a distinct order, and should be considered as an abnormal and passerine development of Rasores, representing the Perchers, but having no direct affinity with them.

Having thus eliminated a considerable number of generally large-sized birds, we have still remaining by far the larger portion, forming the Passeres, Insessores, or Perching birds. Out of about 7000 known birds, upwards of 5000 are Perchers. It is to this great group, or rather to a limited portion of it, that we intend to devote the present paper.

The Passerine order comprises at once the most perfect, the most beautiful, and the most familiar of birds. The feathered inhabitants of our fields, gardens, hedge-rows and houses belong to it. They cheer us with their song, and delight us with their varied colours. Their activity and elegant motions are constant sources of pleasure to every lover of nature. They are the birds with which from our infancy and boyhood we are most familiar, and we therefore involuntarily derive from them that ideal or typical form of animal life with which we connect the general term, Bird. And thus doing, who can doubt but that we are correct? The lightness, activity, elegant forms, brilliant colour and harmonious voice by which birds as a whole are peculiarly distinguished from all other animals, find in this group their fullest expression and most complete development. Here too the greatest variety of forms and habits is found, which are all connected together by such insensible gradations, that to discover in every case their true affinities has ever been and still remains one of the most difficult, and at the same time most interesting problems the naturalist has to solve.

The writer of this paper has enjoyed the great privilege of observing the habits of many tropical birds in a state of nature in S. America, and is at present doing so in the Indian Islands. Every naturalist knows how important this is towards a proper appreciation of the affinities of Birds, to which their habits are generally a sure guide, or at all events of much value in conjunction with other structural characters. Without pretending to any great knowledge of anatomy, he believes that no intelligent person can be in the constant habit of skinning birds without obtaining much information on very important parts of their internal structure. Even mere external characters, such as the texture and arrangement of the feathers, the form and structure of the tarsi, the form of the nostrils and of the tongue, can be examined far better in a recently killed bird than in a dried or mounted specimen. In the process of skinning we also ascertain the thickness and tenacity of the skin, the solidity of the bones, the form and strength of the skull, and the texture and contents of the stomach, which characters are perhaps, for the determination of affinities, of as much importance as any which can be pointed out. Observations of this nature have been applied by him to an arrangement of the Passeres; not as a perfect scheme, but as a starting-point to guide future inquiries. One portion of this arrangement, with the families included in which he is best acquainted, he now wishes to submit to the judgment of ornithologists.

The method illustrated at the commencement of this paper, of marking off certain groups from the general mass, has been satisfactory, because the portions so severed have been not only capable of definition, but have contained only species which have agreed in all essential points of their structure and œconomy. They have therefore met with general acceptance, and in all the different systems of ornithologists these groups have scarcely suffered any variation. But in attempting to carry out this system in a further division of the Passeres, no such satisfactory and generally accepted results have been produced. No systematist has been satisfied with the arrangements of his predecessors, and, after an endless variety of divisions and subdivisions, we are as far off from any generally accepted system of arrangement as ever.

The reason of this we conceive to be, that we have to deal with a mass of species in which the series is so nearly complete, that there are no more of those great chasms separating considerable portions from each other, and that the affinities are so intricate and minutely varied, and so cut up as it were by minor gaps between genera and families, that any attempt to form great and well-marked subdivisions must fail, for the simple reason that such are not marked out by nature. In such a case an *arrangement* may be possible, but a *classification* may not be so. We must therefore give up altogether the principle of *division*, and employ that of *agglutination* or juxtaposition. We shall best explain our meaning by pointing out the errors we conceive to have been produced by the former method in most modern classifications.

The system of Cuvier, as modified by Vigors, Swainson, and G. R. Gray, may be fairly taken as that most generally in use, at least in this country. The tribes of the Conirostres, Dentirostres, Fissirostres, Tenuirostres, and Scansores, are said to be the natural divisions of the Passeres, the main difference of opinion being as to whether the Scansores should or should not form a separate order, a question we believe to be of no importance whatever. These divisions being accepted, every bird is forced into one of them, and the result has been the most incongruous and unnatural combinations. For instance, in the Tenuirostres are combined the Humming Birds and the Sun Birds (*Nectarinia*), families which in a natural arrangement would have, in our opinion, the mass of the other Passeres intervening between them. In the case of these two families, a mere outward resemblance appears to have been universally mistaken for an affinity. A similarity in size, in the prevalence of metallic colours, and in the slenderness of a very variable bill, has been taken to overbalance the most important structural differences. The universal characteristics of the Hummers are, excessively long wings and as excessively small feet, with more or less united toes. They take their food exclusively on the wing. Every motion is made upon the wing. The feet are solely used as means of support, never for locomotion. The Sun Birds and their allies, the *Cærebidæ* of America, have on the other hand long legs and toes, the hinder toe especially being very long and powerful; they are therefore as capable of hopping and perching as any of the most highly developed Passeres. Their wings, too, are short and round, quite incapable of any powerful flight, and their tail almost invariably short and even. Much stress has been laid upon the similar form and structure of the tongue. But the extensile tubular tongue only exists in some genera (*Anthreptes*, *Arachnothera*), while in others it is short, flat, and not extensible (*Dicæum*,

Cinnyris, *Cærebidæ*). There is therefore no general agreement of structure to unite these groups, except the solitary and trivial one of an elongate and slender bill.

On similar principles, we believe the Conirostres and Dentirostres to be equally untenable and unnatural. They are professedly founded on one character only, and not on general structure; and it is therefore not to be wondered at, that in their attempts to pay some little regard to natural affinities, while forcing the genera and families into these divisions, no two naturalists should be able to arrive at the same results. The association of the Hornbills with the Crows, the separation of the Larks from the Wagtails, and the necessity for putting Dentirostral birds (Tanagers and Jays) in the Conirostral tribe, are some of the inconsistent results of the system.

The remaining groups, the Scansores and the Fissirostres, we believe to be much more natural, and in fact to be the only ones which can be distinctly separated from the Passeres, of which they form an abnormal development. It is to the arrangement of these two groups that we more particularly address ourselves.

The Fissirostres are those passerine birds whose feet are adapted solely for a state of rest,--all motion being performed by the wings. With very rare exceptions, they never move the shortest distance by means of their feet,--a character which distinguishes them at once from all other Passeres, which either hop, climb, or walk almost incessantly. Such a peculiar œconomy must evidently depend upon corresponding peculiarities of organization; and it is a remarkable proof of how little importance is the form of the bill alone as an index of affinity, that in this highly natural group we find every form of bill,--conical, toothed, hooked, serrated, spear-shaped, curved, and flat. The external characters which distinguish these birds are, very short and weak legs, long, or at all events powerful wings, and a wide gape. Their characteristic habit is to sit motionless, watching for their prey, to dart after it and seize it on the wing, and to return to their original position to swallow it. The groups which possess these peculiarities in the greatest perfection are the Trogons and the Kingfishers, with which we shall commence our inquiry into the extent of the tribe. We must observe at starting, that many continental ornithologists still place the Trogons among the Climbers, because they have their toes placed two and two, whereas those of the Kingfishers are arranged as in the majority of birds. But this is a point of detail which does not in the least affect the habits, for the toes are in both cases connected together at their basis so as to form a broad sole, giving a firm support to the bird without grasping. In both the leg is equally short and weak, and in both all the habits dependent on the feet are precisely similar. Of how very little importance this change in the position of the toes is, unaccompanied by a change in their form, motion, or mode of connexion with each other, we may judge from the fact of there being species of Kingfishers and of Woodpeckers with only three toes, and which yet have no perceptible difference of habits from the rest of the family. It would be as reasonable (and as unnatural) to withdraw these birds from their respective families and form of them a new three-toed family, as to separate the Kingfishers from the Trogons for the reasons assigned. As an instance how totally unable the Trogons are to use their feet for anything like climbing, we may mention that the Trogons of South America feed principally on fruit, which one would think they would get by climbing or walking after if they could. But no; they take their station on a bare branch, about the middle of the tree, and having fixed their attention on some particular tempting fruit, they dart at it, seize it dexterously on the wing, and return to their original seat. Often, while waiting under a fruit-tree for Chatterers or Pigeons, have we received the first intimation of the presence of a Trogon by the *whir-r-r* of its wings as it darted after a fruit. It is curious that this habit seems confined to the Trogons of America. In the East I have never yet observed it, and in the numerous specimens I have opened nothing has been found but insects. The African Trogons also appear to be wholly insectivorous.

Somewhat intermediate between the Trogons and Kingfishers are the Bee-eaters (*Meropidæ*) and the Jacamars (*Galbulidæ*). The last possess the metallic plumage of the

American Trogons, and habits almost identical, while the bill is an approach to that of a Kingfisher. The Motmots (*Prionitidæ*) are also closely related to the Trogons, and may be considered as an offshoot of them, or of the Bee-eaters, parallel to the Jacamars. The habits of all these birds, and of the small Eastern Kingfishers of the genus *Ceyx*, are almost identical, and we think there can be no reasonable doubt of the very close affinity of these five families. The Rollers (*Coraciadæ*) are the next group whose affinities we have to consider. These birds have for a long time been strangely misplaced among the strong-legged Crows and Grackles, whereas the short legs (with the toes united in some genera), wide gape, insect food captured on the wing, the nest in holes of trees or in the earth, and the colour and form of the eggs, all bring them close to the Bee-eaters, Motmots, and Kingfishers, the two former of which appear to be their most direct affinities. The structure of the skeleton, according to the best observers, confirms this result, which may therefore be considered as well established. We now arrive at the *Capitonidæ*, or Puff Birds of S. America, which, like the Trogons, have been often placed among the climbing birds, from having their toes placed two and two. They are, however, true sedentary birds, with habits exactly analogous to those of the Jacamars, Bee-eaters, and Trogons. Some, like the *Tamatias*, frequent the gloomiest and thickest parts of the forest, where they sit motionless on some low branch, and thence take short flights after insects. Others, like *Monasa* and *Chelidoptera*, frequent more open situations, sitting on bare branches often of dead trees, and take longer flights, which in *Chelidoptera* almost vie with those of Swallows for ease and rapidity. The last-mentioned genera make their nests in holes in sloping ground on the banks of streams,--a habit exceedingly general among Fissirostral birds, but we believe quite unknown among the Scansores. Their nearest affinities seem to be with the *Meropidæ* and *Trogonidæ*, though their large heads and heavy bodies would show some approximation to the Kingfishers.

We now arrive at some birds which have always been associated with the present group, of which, in fact, they appear to be the highest development, but which are nevertheless widely separated from the families we have hitherto been considering. They are the Swallows and Goatsuckers. In these the power of capturing insects on the wing has reached its maximum. The gape is enormously wide, the feet generally very short, and the wings long and powerful. Even between the two there is, however, a considerable hiatus; but no one has ever doubted that they are more nearly allied to each other than to any other birds. The question then remains, to which family of the Fissirostres are either of them allied? Where is the link that connects them? The Swallows appear to us to be farthest removed from any of the birds hitherto placed in the tribe. Their small size, powerful flight, and compact plumage, added to their strongly grasping feet, seem to shut out any direct affinity with them. It is, then, in the *Caprimulgidæ* that we must look to discover the affinity we are in search of; and we believe that the only group to which we can approach them is the *Trogonidæ*. The average size, the excessively thin skin, the mass of downy plumage, the general form, the nature of the food, and, in many species, the mode of capture, all point to an affinity in this direction. The different structure of the feet is the most important character on the opposite side of the question; but as this equally exists in such an undeniably closely allied family as the *Prionitidæ*, it need not be considered an insurmountable obstacle. There is, no doubt, still a very wide chasm to be passed over; but it will be still wider if we compare them with any other family of the Fissirostres, with which their anatomical structure, as well as the general considerations before alluded to, compel us to place them. We consider, then, the Swallows and Goatsuckers to exhibit the greatest development of the Fissirostral form; or, if the term is preferred, to be the typical groups. And, as a consequence of this position, they can neither of them serve as the connecting links or transition to any other tribe or order of birds; for if the Fissirostral character is what serves to distinguish this tribe from all others, it must certainly follow that those birds which have this character in its highest development must be most distinctly separated from all the species of any other grou We have here another reason for believing that the resemblance of the Goatsuckers to the Owls is one of externals, and not of essentials,--of analogy, but not of affinity.

We have now briefly passed in review all the families which possess the characters of Fissirostral birds in a plain and obvious degree, and which, without the greatest violation of their natural affinities, cannot be placed elsewhere in the system; but there are several others which have been associated with these by many naturalists, some of them we believe erroneously. And, first, the *Eurylaimidæ*, or Broadbills of the Eastern Archipelago, have been, and still are, generally placed among the Fissirostres. They have, however, in our opinion, no right whatever to this situation, being true Passeres, allied to the *Cotingas* and *Querulas* of S. America. The *Eurylaimus Javanicus* and the *Cymbirhynchus macrorhynchus* feed in the same manner as ordinary perching birds, hopping about the branches of trees, and picking off the fruit, which forms their principal subsistence. Their legs are of a moderate length, their toes strong, and the hind toe large and powerful, *which is never the case in true Fissirostres.* It is this peculiar conformation of the feet which affects the whole character and habits of these birds, which is not the case with the trivial circumstance of the toes being partially united. They have, moreover, none of the Fissirostral habits, nor any modification of them, as they do not capture insects on the wing, those which we have found in their stomachs being always ants and small Coleoptera, picked off the bark and leaves of the trees. To what erroneous results a dependence on such isolated characters as more or less united toes will lead, is seen by the Manakins and *Rupicolæ* of S. America being also often considered as Fissirostres on the same grounds as the *Eurylaimi*. Those birds are, however, purely frugivorous, are excessively active on their feet, have strong skins, firmly-set plumage (as have also the *Eurylaimi*), and, in fact, not one single natural character which can remove them from the great mass of Perching birds. The beautiful *Calyptomena viridis* has also all the habits of the Chatterers, and cannot be separated from them without a great violation of natural affinities.

The little Todies of the W. Indies have also been usually classed as Fissirostres; but their moderately long and slender legs, short rounded wings, and their excessive activity on their feet, are so totally opposed to the characters of every other member of that group, that we think them far more naturally associated with such Flycatchers as *Todirostrum* and *Megalophus*. From the description of Mr. Gosse, in his 'Birds of Jamaica,' it appears that they are most active little birds, hopping, perching, and flying after insects in every possible place and position: how totally opposed is this to the general character of the Fissirostres, which are sedentary and motionless, except when upon the wing! We cannot allow the one character of their nidification in holes on the banks of streams to counterbalance such a total diversity in structure and habits. It is, besides, impossible to point out any one group of Fissirostres to which they can be said to make any approach, whereas they have the greatest possible resemblance to the genera of Flycatchers above mentioned. We must therefore unhesitatingly decide, that the Todies are not to enter among the Fissirostres.

In place, however, of these two families which we reject, we introduce two others which have not generally found a place here. From an examination of the structure of the feet and toes, and from a consideration of their habits, we are led to conclude that the Hornbills are Fissirostral birds, though of a very abnormal form. Their very short legs, and united toes with a broad flat sole, are exactly similar to those of the Kingfishers. They have powerful wings, but their heavy bodies oblige them to use much exertion in flight, which is not therefore very rapid, though often extended to considerable distances. They are (in the Indian Archipelago at least) entirely frugivorous; and it is curious to observe how their structure modifies their mode of feeding. They are far too heavy to dart after the fruit in the manner of the Trogons; they cannot even fly quickly from branch to branch, picking a fruit here and a fruit there; neither have they strength or agility enough to venture on the more slender branches with the Pigeons and Barbets, but they alight heavily on a branch of considerable thickness, and then, looking cautiously round them, pick off any fruits that may be within their reach, and jerk them down their throat by a motion similar to that used by the Toucans, and which has been erroneously described as throwing the fruit up in the air before swallowing it. When they have gathered all within their reach, they move sideways along the branch by short jumps, or rather a kind of shuffle, and the smaller species even hop across to other branches, when they again gather what is within their reach. When in this way they have progressed as far as the bough will safely carry them, they take a flight to another part of the tree, where they pursue the same course. It thus happens that they soon exhaust all the fruit within their reach; and long after they have left a tree, the Barbets and *Eurylaimi* find abundance of food on the slender branches and extreme twigs. We see, therefore, that their very short legs and syndactyle feet remove them completely from the vicinity of the Toucans, in which the legs are actively employed in moving about after their food. Their wings, too, are as powerful as those of the Toucans are weak, and it is only the great weight of their bodies that prevents them from being capable of rapid and extensive flight. As it is, their strength of wing is shown by the great force with which they beat the air, producing a sound, in the larger species, which can be distinctly heard a mile off, and is even louder than that made by the flight of the great Muscovy Duck. They are still farther removed from the Crows, with which they have also been very generally associated, solely because they are *Conirostres*, or conic-beaked!--another instance of the extremely erroneous results which are arrived at by a dependence on a single character, and especially on one which so little influences the habits of a bird as the external form of its bill.

The preceding deductions from the habits of these birds had been made before I became aware that Mr. Eyton had arrived at similar results from anatomical considerations alone; and I had great pleasure in finding that there was such solid support for the opinion which I had formed, entirely from my own observations. The only question that remains then is, to what family of the Fissirostres do they most nearly approach? A careful consideration leads us to fix upon the Kingfishers. They are among the largest birds in the group, they have the largest bills, and, in the structure of the feet, the two are almost identical. The Hornbills of Africa are said to feed principally on reptiles, as do the King-Hunters (*Dacelo*) of Australia. We look upon Hornbills, therefore, as one of the abnormal developments of Fissirostral birds, of which they are the largest, the least elegant, and the least gifted with facilities for locomotion and for obtaining their food; and that their nearest affinities lie in the direction of the Kingfishers.

The remaining family, which, according to our views,

belongs to the Fissirostral tribe, is that of the *Trochilidæ*, or Humming Birds, hitherto always placed in the Tenuirostres, which we have before adduced some reasons for believing to be an altogether artificial grou In this innovation we are not aware of having any support; yet we think it possible to show good reasons for it. What is it that characterizes the Fissirostral group but minimized feet and maximized wings, always connected with some modification of structure, adapted to give facilities for seizing the food with the mouth? and all these the Hummers possess in a remarkable degree. In the ease and flexibility of their motions on the wing they surpass even the Swallows. Their little feet exactly resemble those of some of the Swifts (*Dendrochelidon*), and the long, variously-curved bill and extensile tongue give them the same facilities for obtaining their food as do the short bill and wide gape in the other. They are, too (we believe), like the Swallows, purely insectivorous; for in every specimen we have examined the stomach was full of insects, principally minute flower-frequenting Coleoptera. No doubt they do partake occasionally of the nectar, or the pollen of the flowers they frequent, but as a delicacy rather than as solid food. The firmness and solidity of their muscle, the thickness of their skin, and the immense muscular exertion which they constantly make, can doubtless be supported only by animal food, which the very small space occupied by their stomach and intestines also shows to be the case. But a stronger proof of this assertion is, that there are many species which never frequent flowers! All the species of the genus *Phaëthornis* which we met with on the Amazon were found only in the lower parts of the forest, among the shrubs and palms which rise only a few feet from the ground. Here we have often seen them searching the leaves for insects, supporting themselves almost motionless in the air, their body erect and their bill pointed upwards, and passing rapidly over the under surface of each leaf in succession. They would often dart suddenly out into an open space, and remain motionless a few feet from my face, and then fly off again to repeat their search for their favourite food. At the various trees in flower about which I have observed and shot Hummers, I have never seen a species of *Phaëthornis*, whereas I used daily to meet with them in the gloomy jungle, where not a flower exists; and yet, from never being able to see them at a greater distance than three or four yards, they were the most difficult of all to obtain, without blowing them to pieces. Many species also hunt for insects in the air, exactly like true Fissirostral birds. I have often observed them in the evening, on the banks of streams, coming out of the jungle just as the Goatsuckers were beginning to appear, and darting about after the mosquitos and other minute insects, returning after each short circuit to the edge of the forest, where they remained balanced in the air for a moment and then darted off again. At other times they will sit on the topmost twig of a dead tree, and making short circuits in the air, return to it, exactly in the same manner as do the Jacamars and Puff Birds.

There is also another interesting fact to be mentioned. I had brought me a nest containing two little Hummers, apparently very recently hatched. I tried to feed them, and gave them, first, according to established rule, syrup made of honey and water, and also of molasses; but the poor little creatures did not at all like it, though they opened their mouths as if ravenously hungry. They were nearly choked by the liquid, and tried all they could to spit it out, which they generally succeeded in doing. Finding all my efforts to suit their taste in vain, I resolved to try if they liked insects better, and caught some minute flies which were very abundant. On dropping one of these into their mouths, they immediately closed their beak and by a great muscular effort of the throat swallowed it, and opened their mouths again for another. In this way they would each take fifteen or twenty little flies one after the other before they were satisfied. I thus kept them alive three or four days, and could I have bestowed sufficient time and constant attention upon them, there is no doubt they would have lived much longer. At all events the experiment satisfied me that the young Hummers are fed by their parents with insects, and not with honey.

I also observed that the beak of these little birds was very short, triangular, and very broad at the base,--in fact exactly the beak of a Swallow slightly lengthened. We see therefore in the Humming Birds an extreme and peculiar development of the Hirundine form of the Fissirostres. The wings of the Swallow, already among the longest in the whole class of Birds, are still further lengthened. The feet, already so short, are still further reduced. The plumage, which in the Swallow has become more compact and appressed than in any other of the Fissirostres, has these qualities still further developed. The skin, which in the tribe generally is thin and tender, is in the Swallows comparatively thick and strong, and in the Hummers is perhaps stronger in proportion to their size than in any other birds. The bill is that of a Swallow lengthened out to contain the long and extensile tongue; and the vital force and energy which enables the Swallows to enjoy such long-continued and rapid motion, seems here to have reached a point beyond which further development is scarcely possible. How then can we refuse them a place among those birds of which they possess the distinctive characters in the most eminent degree, while at the same time we keep together birds as different from each other as the Kingfisher from the Swallow, because they possess those characters?

But it will be objected that the structure of the tongue is so different and peculiar, and agrees so well with that of the Sun Birds. But we have already mentioned, and again repeat, that in closely allied genera of Sun Birds the tongue is totally different, and that therefore it is not a character to outweigh the whole structure and habits of a group of birds: moreover, in other groups the same difference in the tongue is not held to be sufficient to separate birds otherwise allied. The Cuckoos and the Toucans have ever been placed near each other, yet how different their tongues! while the Woodpeckers, still farther differing from them, are notwithstanding placed in the same tribe of Climbing birds. We might also expect, that when the structure of a bird had become so peculiarly modified as to bring it to seek the same food in the same places as another bird of quite a different type of structure, we should find each of them gifted with the same peculiarly modified organ adapted to such habits; and we therefore find that the Sun Birds and the Hummers, though with a widely different general structure, yet have a similarly constructed tongue which they both use in extracting minute insects from flowers and leaves. An exactly analogous instance exists in the *Picidæ* and *Dendrocolaptidæ*, two families as different in general structure as the *Trochilidæ* and *Nectarinidæ*, but which yet have one striking similarity in the rigid tail, which enables them both to rest vertically against a tree while extracting insects from the bark. We contend that these cases are strictly analogous, and that there is no more real affinity in the one case than in the other. The character which is most opposed to this view of their affinities is their nidification; but we think this is not only not an insuperable obstacle to their being thus placed, but one that we might to some degree have anticipated. In some of the Swallows we have already seen one deviation from the general character of the tribe in the carefully constructed nest of clay. In the Goatsuckers we have another, the Podargus of Australia forming a nest of sticks and grass on the branches of a tree. We should therefore expect that birds so peculiarly and highly organized as the Hummers, so aërial in their habits, and so intimately associated with flowers and foliage, would have a modified and characteristic form of nidification.

We have now completed a very brief and imperfect review of those families which we conceive can be separated from the mass of Passeres to form the tribe of the Fissirostres, and in the accompanying diagram we have endeavoured to represent at one glance their various affinities.

Diagram of the Affinities of the Fissirostres.

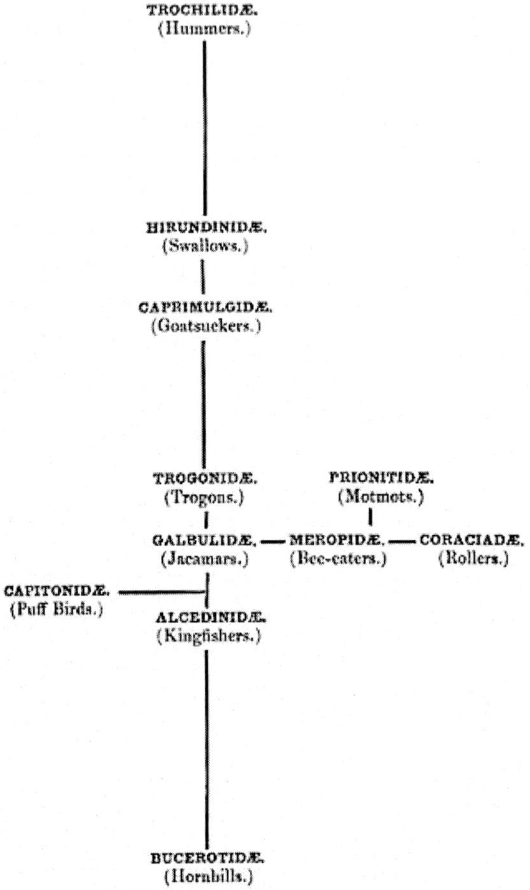

It is intended that the distances between the several names should show to some extent the relative amount of affinity existing between them; and the connecting lines show in what direction the affinities are supposed to lie.

By referring to the diagram it will be seen that there are seven families placed close together, forming a central mass. Beyond the Trogons at some distance come the Goatsuckers and Swallows, while at the greatest possible distance from each other are the Hornbills and the Hummers, the former having a distant affinity to the Kingfishers, the latter to the Swallows.

We may here mention that it is an article of our zoological faith, that all gaps between species, genera, or larger groups are the result of the extinction of species during former epochs of the world's history, and we believe this view will enable us more justly to appreciate the correctness of our arrangement. For instance, let us suppose that the gaps shown in this diagram have been all filled up by genera and families forming a natural transition from one of our groups to the other, and we shall be able to judge whether our arrangement will agree with such a supposition. Thus, if the space between the Kingfishers and Hornbills has been filled up by a natural succession of families, we can see that the change must have been to heavier, larger, and larger-billed birds, and we see such a change begun already from the Jacamars to the Kingfishers. So from the Goatsuckers to the Swallows the change is to smaller, stronger-winged, thicker-skinned, and brighter-coloured birds,--exactly the kind of change which continued on will lead us to the Hummers. On the same principles we may conclude, that as the change from the Jacamars to the Rollers is to stronger-legged birds which do not feed solely on the wing but also on the ground, so the same change continued on would lead us to true Passeres in which the peculiar Fissirostral characters altogether disappear. The *Coraciadæ*, therefore, are either an extreme Passerine form of the tribe, or else form a transition by direct affinity to the Passeres.

The method of representing affinities here adopted we believe to be of the highest value. It is founded on the method suggested by the late Mr. Strickland, and which we believe Dr. Lindley has been the first naturalist to adopt, namely that of placing to the right and left of every family or other group the names of those to which it is most nearly allied. But this alone conveys no idea to the mind, especially in an extensive group, till represented by a diagram, the most convenient way to construct which we have found to be as follows. Suppose you have a family of a dozen genera which you wish to arrange; first write down the names of the genera in any order, and right and left of them those to which you believe them to be allied most closely. Then take a dozen pieces of paper or card cut out with a gun-wadding punch, and on each write a name of a genus. Place them on a table and arrange them according to your list. This will not always be so easy a matter as it appears at first sight, for you will most likely find that you have set down some conflicting affinities, or that you have mistaken some mere analogies for affinities. When you have them in tolerable order, the next thing is to get the distances between them to bear some proportion to the closeness or remoteness of the affinities, and lastly, bring the whole into symmetry by placing what appears to be the main line of affinities in a straight line, and bringing the others into branches right or left from it. When this is done, the positions can be copied on a sheet of paper and kept for reference as a trial-arrangement, which is to be tested by every new fact that is procured, and by any additional knowledge that may be gained on the structure or habits of any of the species. The advantage claimed for this particular form of diagram is that it can be printed with ordinary type, whereas any circles or figures to represent the groups require woodcuts or lithographs. It is much to be wished that in every systematic work each tribe and family should be illustrated by some such diagram, without which it is often impossible to tell whether two families follow each other because the author thinks them allied, or merely because the exigencies of a consecutive series compel him so to place them. Thus, Bonaparte places in his 'Conspectus,' the *Trochilidæ* between the *Cypselidæ* and the *Phytotomidæ*. By making them follow the Swifts he would seem to take the same view of their affinities as is here done, but by placing immediately after them the *Phytotomidæ*, one is at a loss to understand by what principles he has been guided. An explanatory diagram, or even the plan of denoting the affinities as adopted by Dr. Lindley, would remove such doubts, and render a work of such great labour and research as the one referred to less likely to be misunderstood.

On the Affinities and Limits of the Scansorial Birds.

However much systematists have differed as to what families should enter into or be excluded from the Scansores or Climbers, considered as a natural group of Birds, there are four families which have formed part of it in every system. These are the Woodpeckers (*Picidæ*), the Parrots (*Psittacidæ*), the Cuckoos (*Cuculidæ*) and the Toucans (*Rhamphastidæ*). We may therefore take these as a basis, to inquire in what respects the Scansores differ from the true Passeres and from the Fissirostres, and to deduce their natural characters. Having done this, we may inquire further if any and what other groups can be naturally associated with them.

The first thing that strikes us on comparing these birds with each other is, that we could hardly pick four families from the whole class which should have more diversified forms of bill. The resemblance and affinity between them must therefore exist in other parts of their body, and we find it in their wings and feet. The former are generally short, rounded, and very weak, quite incapable of rapid or long-continued flight, while the latter are remarkably large, powerful, and peculiarly formed. They may be said therefore to be the very reverse of the Fissirostres, whose grand features are large wings and small feet, while in the Scansores the small wings and large feet are equally characteristic. But it is the peculiar structure quite as much as the size of the feet to which we must pay attention. The toes are always exceedingly long, and the outer toe is either turned completely backwards or nearly at right angles to the others. This toe is often the longest of all, while the true hind toe is always small, and sometimes altogether wanting. It is this peculiar structure that altogether separates this group from all the short-winged and strong-footed Passeres, whether they are walkers, perchers, or climbers.

The habits that result from this form of foot and wing are, as might be expected, to a great extent characteristic, and will serve us as a valuable guide in those cases of anomalous form and structure where the position of a genus or family might be otherwise doubtful. These birds then are truly arboreal, rarely descending voluntarily to the ground. They use their wings only for passing from tree to tree, and, whether frugivorous or insectivorous, they obtain their food in or upon trees. Their motions along the trunk or branches, or among the thick foliage, are either true climbing, or a succession of rapid hops producing an appearance of climbing. The Woodpecker runs up the vertical trunk, and assisted by a peculiarly modified tail and a powerful wedge-shaped bill, seeks his food beneath the bark. The Parrot climbs, assisted by his hooked bill, after the fruit, which alone he feeds on. The insect food of the Cuckoos is sought for upon the leaves and smaller branches, and they progress among these so rapidly, that they have been constantly mistaken by us for squirrels or other small arboreal animals. The Toucans again hop actively about the tops of lofty trees, devouring an immense quantity of fruit.

Now, though these four families have evidently more connexion with each other than with any other birds, yet they present so many important points of difference, as to show that they are in reality very distant from each other, and that an immense variety of forms must have intervened to have filled up the chasms, and formed a complete series presenting a gradual transition from one to the other. The differences in the form of the bill have already been alluded to, but those of the tongue are perhaps still more extraordinary; the fleshy tongue of the Parrot, the barbed extensile spear of the Woodpecker, the short horny tongue of the Cuckoo, and the long and slender feathered tongue of the Toucan, would seem rather to belong to birds most remote from each other, than to those for whom we can find no nearer allies. We should be inclined to consider therefore that they form widely distant portions of a vast group, once perhaps as extensive and varied as the whole of the existing Passeres.

Notwithstanding the difference of their food, it is evident that the Cuckoos and the Toucans approach more closely to each other than to the others. Their legs are longer, and they consequently hop, which the other two never do. Their bills are similar in form, their plumage is in both much more loose than either in the Parrots or the Woodpeckers, which again, in these peculiarities in which they agree, to some extent approach each other. We would place therefore the Parrots and the Woodpeckers at one extreme of the group, and also considerably removed from each other, while the Toucans and Cuckoos, rather nearer together, should be placed at the other extreme.

The Barbets (*Bucconidæ* of Lesson and Bonaparte, *Capitoninæ* of G. R. Gray) have also been always included amongst the Climbers, but their place has been so often varied and their affinities so much misunderstood that they require a separate consideration, especially as in the systems of Swainson and Gray they have been considered as a subfamily of *Picidæ*, and have therefore not appeared among the families of the Scansores. The only ground for placing them with the Woodpeckers appears to have been that some African species do cling against and peck at trees something in the manner of those birds. Their whole structure however is totally opposed to their being thus placed. In their feet, wings, and the form of the whole body they much more nearly resemble the Toucans. The texture of their feathers, their broad, angular and weak skulls also resemble them, and are strikingly dissimilar to those of the Woodpeckers. From my own observations too, I can assert that, in the habits both of the South American and of the Eastern species, they resemble the Toucans more closely than any other birds; and Le Vaillant makes the same observation with regard to the African species. Besides, the grand characteristic of the Woodpeckers, the barbed and extensile tongue, which exists equally in the *Yunx* and *Picumnus*, is totally absent in the Barbets, while their bill is of quite a different type of form, much more nearly approximating to that of the Cuckoos. In their habits too they are equally distinct: they hop and cling, but never climb, and they live almost exclusively on soft fruits. We must therefore consider them as a distinct family, and place them in the vicinity of the Toucans and Cuckoos.

We will now proceed to the consideration of those groups, about the propriety of including which in the Scansorial tribe considerable difference of opinion has existed. These are the Turacos (*Musophagidæ*), the isolated genus *Opisthocomus*, and the *Certhiadæ* including the *Dendrocolaptidæ*. These last have, we believe, only been so placed by Messrs. Vigors and Swainson, but as their classification claims to be pre-eminently "The Natural System," and as it has still some advocates, it deserves to be carefully examined. What are the characters then by which the *Dendrocolaptidæ* are supposed to be united to the *Picidæ*? They appear to be these: both are true Climbers, both have a rigid tail which assists them in maintaining an erect position, and both feed on insects which they obtain upon the trunks and branches of trees. On the other hand, they present many and important differences. The long, slender, curved bill and short, horny, non-extensile tongue of the Creepers are very far removed from the strong straight bill and extensile barbed tongue of the Woodpeckers. But this, it may be said, is of no importance, as a similar difference exists in the other families admitted into the Scansores. This is true; but then those birds agree in having the same form of feet, which is of far more importance in this case than it may at first sight appear, for we shall be able to show not only that the Creeper's foot is very different from the Climber's, but that *it is further removed from it than is that of any other of the Passeres.*

The characteristic form of foot in the *Certhiadæ* and *Dendrocolaptidæ* is to have the toes placed normally, three forward and one backward, and to have the forward toes all connected together at their bases, particularly the outer toe, which is generally longer than the inner, and often connected to the middle toe as far as the second joint. The result of this conformation is, that the forward toes do not spread much laterally, but form one line of support opposed to the hind toe. This hind toe also is remarkably long and powerful, and armed with an equally powerful claw. This peculiar structure has been gradually arrived at, through the most nearly allied families of Passeres. Passing from the Wagtails and Larks through the *Anabatidæ* to the *Certhiadæ*, *Sittidæ*, and

Dendrocolaptidæ, we find the outer toe gradually more and more united to the middle one, and the hind toe becoming gradually larger and more developed; so that we are justified in asserting that we see here that peculiar modification of the normal (3/1-toed) foot which is adapted for climbing. In the Woodpeckers, however, we find the outer toe always turned completely backwards, and therefore quite free from the middle toe. The true hind toe is also invariably small and weak, and of so little importance that in several species of Woodpeckers it is altogether wanting, without at all diminishing the bird's powers of climbing. If we compare this foot with that of the other Scansores, we shall find that it is an extreme modification of the Scansorial form, adapted for true climbing. The toes are all more powerful, the claws much stronger, the outer toe longer, and the hind toe smaller. If therefore the structure of foot in the Cuckoos and Turacos, where the outer toe can be placed either forward or backward and the hind toe remains moderately developed, is (as is universally allowed) the link between the 2/2-toed and 3/1-toed form, then it follows that of all 2/2-toed feet the Woodpecker's is most removed from the 3/1-toed, and of all 3/1-toed feet the Creeper's is the furthest removed from the 2/2-toed;--and thence as a further deduction it follows, that the feet of the Creepers and Woodpeckers are the furthest possible removed from each other. When, in addition to this remarkable result, we consider that the structure of the climbing tail is totally dissimilar in the two cases, we shall see that there exist no grounds whatever for establishing an affinity between the two families, and that the Creepers must not only be separated from the Scansores, but in a natural arrangement will be placed at a very considerable distance from them.

The *Musophagidæ*, containing the Turacos and Plantain-eaters, have been placed among the Scansores by the continental ornithologists (Temminck and Vieillot), while in England they have been considered to be Conirostres by Swainson and Gray. We believe the former are correct; for these birds have the short, rounded and weak wings of the Cuckoos and Toucans, and consequently very imperfect flight, while their legs are very strong, the outer toe long and versatile, but rather less so than in the Cuckoos, and the hind toe, as in all Scansores, short. Their habits are described as being almost exactly those of the Toucans and Barbets, their plumage is of a similar texture, while the short crest at the back of the head is similar to that of the Woodpeckers. Some species are said to be able to cling to vertical trunks. Their internal structure and the form of the sternum appear to correspond exactly to this view of their affinities, which is still further confirmed by their nidification, like that of all other Scansores, in hollow trees, so that they may be well placed in the wide interval between the Cuckoos and Toucans on the one side, and the Woodpeckers and Parrots on the other, but rather nearer to the former than the latter. If, on the other hand, we place this group among the Conirostres, we can give no such satisfactory account of its structural affinities. Swainson places it between the *Fringillidæ* and the *Buceridæ*. The former have all well-formed Passerine feet, the hind toe always well developed, and the outer toe never so long as the middle one; they have generally powerful wings, and are of such a uniformly small size as not even to give them an appearance of affinity with the Turacos and *Musophagidæ*. The Hornbills are if possible still further removed, as our previous account of their habits and the structure of their feet will at once show. We cannot believe that so very acute and observant a naturalist as Mr. Swainson could have been led to propound these as natural affinities, had he not been blinded by his belief in the universal existence in nature of a numerical and circular arrangement, which, without disproving it in any particular cases, we believe can be shown to be absolutely untenable on two general grounds. 1st. Geological investigations prove that the animals now existing on the earth are probably not one-tenth, perhaps not one-hundredth, of those which have existed; for all before the Tertiary epoch were of different species and mostly of different genera, and thousands of other genera, families, and whole orders must have existed of which we are absolutely in ignorance. If therefore this regular system were true of the whole, it must be quite imperceptible in the mere fragment we have an acquaintance with. Instead of complete circles being the rule, they should scarcely ever exist; in fact, the gaps left in the system by its authors do not leave room enough for all the forms that must have become extinct. 2ndly. This system absolutely places limits to the variety and extent of creation; for it is said that every group can only contain five subgroups, and the number of gradations of groups is fixed. For instance, in a family there can be only five subfamilies, in each of which there can be only five true genera, and again in each genus five subgenera. In the *Psittacidæ* therefore there can be but twenty-five generic forms, and when those are all known, not only is it declared to be impossible to discover a new one, but it is also asserted that no others can possibly ever have existed and become extinct. This is the logical deduction from any system of definite numbers in natural history, and it is one that should convince every person of the false basis on which all such systems rest.

Having determined the position of the Turacos, we shall next have to consider that remarkable bird, the *Opisthocomus cristatus*. This has been and still is placed among the Gallinaceæ by most continental authors. Mr. G. R. Gray, however, places it near the Turacos, and Mr. Swainson in the family of the Cuckoos. We believe it should be placed between the two, or rather as a lateral branch from the Turacos. This bird is very abundant on the banks of the Amazon, where we have often observed and shot it. It frequents low bushes on the river's edge, where it feeds on leaves, principally those of a gigantic Arum. It never goes on the ground. This circumstance, combined with the fact of its having no gizzard, would at once decide that it is not Gallinaceous. Our own impression at the time, from its general appearance, flight, and habits, was, that it was a gigantic Cuckoo. Its long crest remarkably resembles that of the genus *Diplopterus*, several species of which occur in the same district, and they both have the habit of throwing it up when alarmed in exactly the same manner. In its bill and general form it approaches the Turacos more nearly than any other bird. The only difficulty is in the feet, which, though similar in form, have not the versatile outer toe of those birds. This however seems of less importance, because a genus of *Musophagidæ* (*Schizorhis*) has also all the toes directed forwards. The short wings, weak heavy flight, strong legs, long toes, and the character of the plumage, added to the resemblances already pointed out, certainly justify us in believing this to be the true position for this singular bird, while its peculiar food and internal structure show that it is to some extent isolated, and cannot be referred to any known family.

We have now only one more group to introduce into our Scansores, but it is one of extreme interest, as tending in some degree to fill up the wide chasm which separates the *Psittacidæ* from all other birds. This we believe is done by the *Coliidæ*, a small group of birds peculiar to Africa, and which have been generally classed as Finches, from their small size and thick beaks. The particulars which Le Vaillant gives of their habits are however exceedingly curious, and show a resemblance to the Parrots which no other birds exhibit. They live entirely on fruits, never touching either seeds or insects. They never perch or jum They walk with the whole tarsus applied to the ground, creeping as it were upon their belly. They are very fleshy, and weigh twice as much as another bird of apparently the same size, for their feathers are so short and so close-laid upon their body, that they are really much larger than they appear. They have also very weak wings, and can fly a very short distance. They climb up to the top of a tree or bush to fly to another, and in doing so, lose elevation so as generally to arrive at the foot of it. They climb one foot after the other, and *help themselves on with their beaks.*

Now, almost the whole of this description will apply to some of the Parrot tribe and to no other birds. Their bill is an approach to that of the Parrot; the upper mandible being thick, much curved, and acutely pointed, while the lower is much smaller and nearly straight,--a form quite different from that of the Finches. The feet are very peculiar, the hind toe being small and capable of being turned forward. The tongue is described as cartilaginous and flat,--one step from the ordinary horny-tipped tongue to the fleshy one of the *Psittacidæ*. We consider therefore the *Coliidæ* to be more nearly allied to the Parrots than any other birds, and to be an isolated link serving to connect them with the other Scansores in the direction of the *Musophagidæ*.

In the accompanying diagram we have endeavoured to lay down the families of Scansorial birds, so as to represent their respective affinities; but the very imperfect and fragmentary state in which according to our views the group exists, prevents our arriving at a very satisfactory result.

We may here remark, that we can never hope to arrive at the true direction and amount of the affinities of the several families of birds, owing to our complete ignorance of the extinct forms. It is probable that in very few cases is there a direct affinity between two groups, each being more or less distantly related to some common extinct group, so that we should represent their connexion more accurately by making our central line a blank, for the extinct portion of the group, and placing our families right and left, at different distances from it. We should thus see the reason why we so rarely find one family or genus exactly intermediate between two others. For instance, though the Cuckoos are by their feet intermediate between the Turacos and Toucans, yet their different plumage and their insect food show that they are more properly a lateral branch from some common central group now extinct.

Having thus determined the extent of the two groups

which can be separated from the Passerine birds, there remains an extensive series of species which we believe constitute one great group of equal value with those we have already defined. This group may be called the normal or typical Passeres, and consists of about thirty-five families, containing between three and four thousand species, or at least half of the known birds. These we believe are too intimately connected with each other to allow of their being separated into a few great divisions without violating many of their natural relations. They have all normal or 3/1-toed feet, which are never so short or weak as to be unadapted for progression. The bill is always moderate in size and form, and in the few cases where it is peculiarly modified, as in some species of *Dendrocolaptes*, other species in the same family possess the normal form. There is also a remarkable moderation in size; for though the species are so numerous, there are none either so large or so small as are to be found in the two abnormal groups. There is also a much greater uniformity in texture of plumage

Diagram of the Affinities of the Scansores.

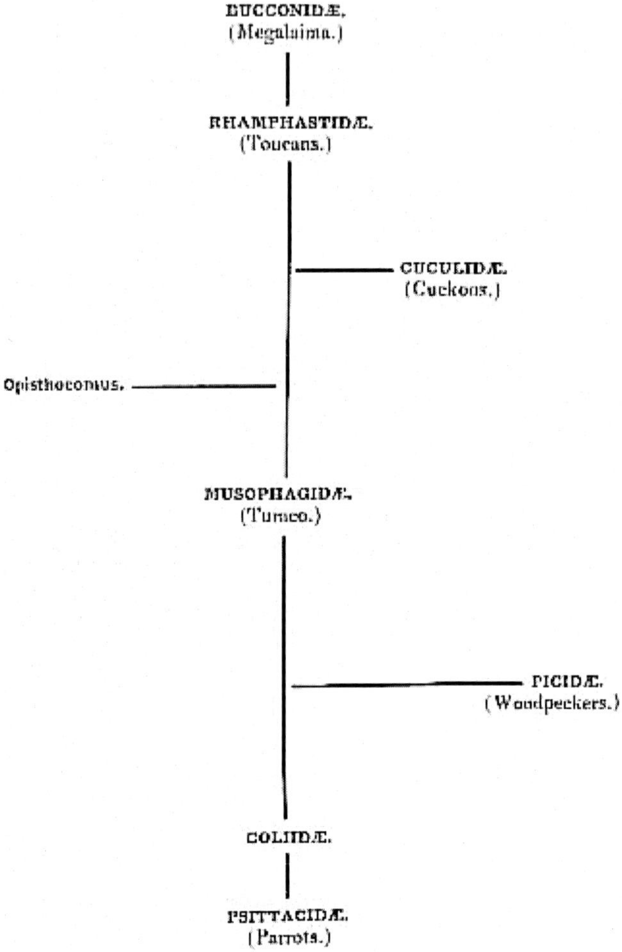

and in form, as well as in habits, which binds the whole into one compact and natural grou It is also a most important point to consider that there are no isolated families,--none but have numerous points of connexion and transition with others; and to such an extent is this the case, that there is scarcely an extensive family group about the limits of which ornithologists can agree. The Thrushes, Warblers, Flycatchers, Chatterers, Tanagers, Finches, Shrikes, Bush-Shrikes, and many others are in this condition, and offer a striking contrast to the families of the Fissirostres and Scansores, about the limits of every one of which there is scarcely any doubt or disagreement whatever. Here then we have three groups, one of which, though very much more extensive than the others, offers less variation in the form and size of the species, and in the modifications of their principal organs. Correct principles of classification would surely oblige us to consider the three groups of only equal rank.

But all the families which compose this group are so intimately connected with each other, that the limits of a great many of them cannot be determined, and there is no family of any extent which does not gradually blend into others. How then can we hope to form two or three primary divisions which shall be sufficiently well marked out to command general acceptance? without some probability of which, the mere multiplication of systems of classification is a nuisance.

We conceive therefore that the efforts of ornithologists should be directed to the study of the different families individually, in order to determine their extent and to point out their true affinities with other families. When this has been done for all, we may be able to arrange the whole group so as to present to the eye a view of the relations of the several parts, and then, and then only, shall we be able to determine whether any and what subdivisions can be established.

There is one other point on which it is necessary to say a few words before concluding this paper. It is on the connexion of the three groups we have here endeavoured to establish with each other. The subject is a most difficult one, and we have been able to come to no satisfactory conclusion upon it. We are inclined however to imagine, that the Puff Birds and Barbets, as exhibiting the least development of the peculiar characters of their respective tribes, may show the line of connexion between the Fissirostres and Scansores, while the Rollers may connect the former with the normal Passeres somewhere near the *Eurylaimidæ*. But a minute and careful examination of the families in question is requisite to decide so nice a point, on which too the greatest light may be thrown by anatomical observations. It is to be hoped that some ornithologist will be found to investigate it fully.

Made in United States
North Haven, CT
04 November 2026

21765292R00030